Cutest Reptiles and Amphibians

Jay Lightsey

aka Outdoors with Jay

www.IntellectPublishing.com

ISBN: 978-1-961485-61-7

Illustrated by Taylor Dinkelman

I am a South African painter. I paint all kinds of animals and people. I love animals but painting is my real passion.

Introduction

My name Is Jay Lightsey aka Outdoors with Jay. I am a teenager now, but I started The Reptile Bus when I was 9 years old. It is a bus that with the help of family and friends I take around my area to spread my passion for these great creatures.

I love educating people about these magnificent reptiles and amphibians.

I have been keeping reptiles since the age of three and enjoy showing my love of reptiles and amphibians on all social media platforms as Outdoors with Jay.

Your Guide Alexander

My name is Alexander, "The Axolotl." Join me to explore some of the cutest reptiles on Earth. As we go, you will learn some fun facts about these reptiles and their habitats.

First, we'll slither our way over to the snakes: the Hognose, Worm Snake, and Sand Boa.

Let's take a closer look at the **Eastern Hognose**. You can see how the Eastern Hognose is cute with its snoot. You might think it's a pretty feature, but that's not the case –it's meant for digging up toads! Its three defense mechanisms are to flatten their bodies, hiss at their predators, and play dead. These snakes are rear-fanged venomous. Not harmful to humans! If you want to find an Eastern Hognose snake, look no further than the Unites States of America!

Next, we are here with the **Worm Snake**. The worm snake eats worms and slugs. Just look at its cute face! As a baby, it is the size of a penny, and the adults are only 12 inches long. It also lives in the United States.

Let's get to this guy, the **Sand Boa**, who spends most of his time under the sand. His cute feature is his eyes, which are almost on top of his head. They are meant for ambushing predators. They raise their heads, and their eyes poke out of the sand and they wait for its prey to come to them. Sand Boas live in Africa.

Up Next, let's crawl our way to the lizards.

The **Red-Eyed Crocodile Skink's** cute features include its red eyes and its dragon-like appearance. They like to live in packs and take care of their young. You can find these cute lizards in Papua New Guinea!

The **Knob-Tailed Gecko** has a cute tail that looks exactly like a doorknob! They have bumps all over their body, which make them unique! You can spot this cute lizard in Australia.

The **Crested Gecko** has long eyelashes and sticky feet. These lizards can easily break off their tails to distract predators. Unlike other lizards, they cannot grow back their tails. Some people call them Frog Butts, because their tailless bottoms look like a frog. You can find this cute lizard in Southern Caledonia.

Now, we will slide our way over to the turtles and tortoises.

This turtle is called a **Fly River Turtle,** also known as the Pig-Nose Turtle. Its cute features are its sea turtle-like flippers and pig nose! You can find this adorable turtle in Northern Australia and Papua New Guinea.

Let's look at the **Red-footed Tortoise**. Its cute feature is its red feet. Red-footed tortoises are omnivores, eating bugs, plants, and rats. These creatures are found in South America.

Next is the awesome **Mata Mata Turtle**. Its cute features include looking like a leaf and having a big smile! A Mata Mata eats fish and small rodents. You can find these cute turtles in South America also.

Let's swim our way over to the next group, the Crocodilians!

Watch out for the **Cuban Crocodile**! It's cute feature is that it gallops, like a horse. It eats turtles, fish, and rodents. This incredible crocodile can be found in Cuba!

Don’t get caught by this slender mouthed **Gharial**, also known as the fish-eating crocodile. Its cute feature is its slender snout, which it uses to catch fish. They live in northern India.

Keep your eye out for the smallest alligator in the world, the **Chinese Alligator**. Its cute features are its black eyes and a big, sweet smile. This alligator is an endangered species with only 150 found in the wild, but their cousin, the American Alligator, is not endangered. They live in China.

Let's hop our way over to the frogs!

First up is the cutest of them all, the **Rain Frog**. Its cute features are that it makes a sound that sounds like a "meep" and has a human-like butt. They only come out during the rain, which is why it is called the Rain Frog. You can find these in Namibia and South Africa.

Milk Frogs! These cute animals are grayish blue with black stripes. When they feel scared, they produce a sticky white substance that looks like milk. These creatures can be found in the Amazon Rainforest.

The last of our cute frogs is the **Pac Man Frog**. This frog has cute horns, eats bugs, rats, small mammals, birds, and each other! They can be found throughout most of South America.

Let's slime our way to the last category: Salamanders and Newts

The **Japanese Fire-Bellied Newt** has a cute red belly. It lives in creeks, rice paddies, roadside ditches, ponds, puddles, and other slow-moving streams in Japan. Watch out! Wash your hands after touching these toxic creatures.

Marbled Salamander. Its cute feature is its black and gray pattern that looks like marble. You can find these cute salamanders in the Eastern United States.

Now for the showstopper, ME! I'm the **Axolotl**. I have pink gills and love to be in the cold water. I used to live in Mexico, but I am critically endangered in the wild. Because of that you can only find me in zoos and pet stores. I stay in my tadpole form for my entire life, unlike other salamanders. I'm incredible because if I were to lose a limb or part of my brain, I can regenerate it.

Thank you for coming on this journey with me through all these cute reptiles and amphibians.

Jay Lightsey

Made in the USA
Monee, IL
05 November 2026

67387952R00017